Module 6 Organic che and analysis

Aromatic compounds, carbonyls and acids

Aromatic compounds

Benzene and aromatic compounds

You should be aware of the terms 'aliphatic', 'alicyclic' and 'aromatic' and be able to apply the IUPAC rules for systematically naming organic compounds. The bonding in benzene, C_6H_6, can be described using both the Kekulé and the delocalised models.

1 a **Calculate the percentage by mass of carbon in benzene. (AO2)** 1 mark

....

....

....

....

b **Describe, with the aid of a suitably labelled diagram, the structure and bonding in benzene using: (AO1)**

i **the Kekulé model** 2 marks

....

....

....

....

ii **the delocalised model** 2 marks

....

....

....

....

c i **State the bond angles in benzene. (AO1)** 1 mark

....

ii **Explain the difference between a σ-bond and a π-bond. (AO1, AO2)** 2 marks

....

....

....

....

Electrophilic substitution and phenols

Benzene and phenol undergo electrophilic substitution reactions, whilst cyclohexene also reacts with electrophiles but the mechanism is electrophilic addition.

The position of substitution on the ring is influenced by groups attached to the ring such as $-OH$, $-NH_2$ and $-NO_2$.

2 Complete the reaction scheme below. (AO2) 5 marks

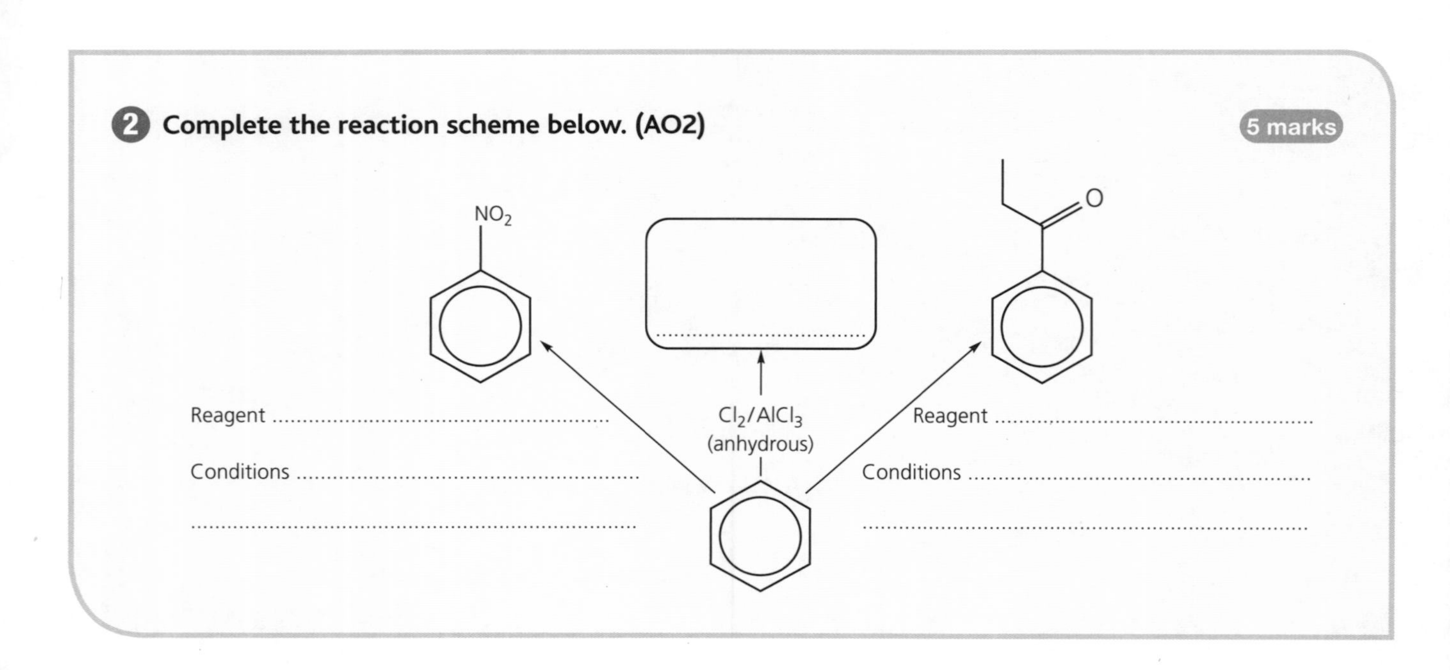

OCR

A LEVEL YEAR 2

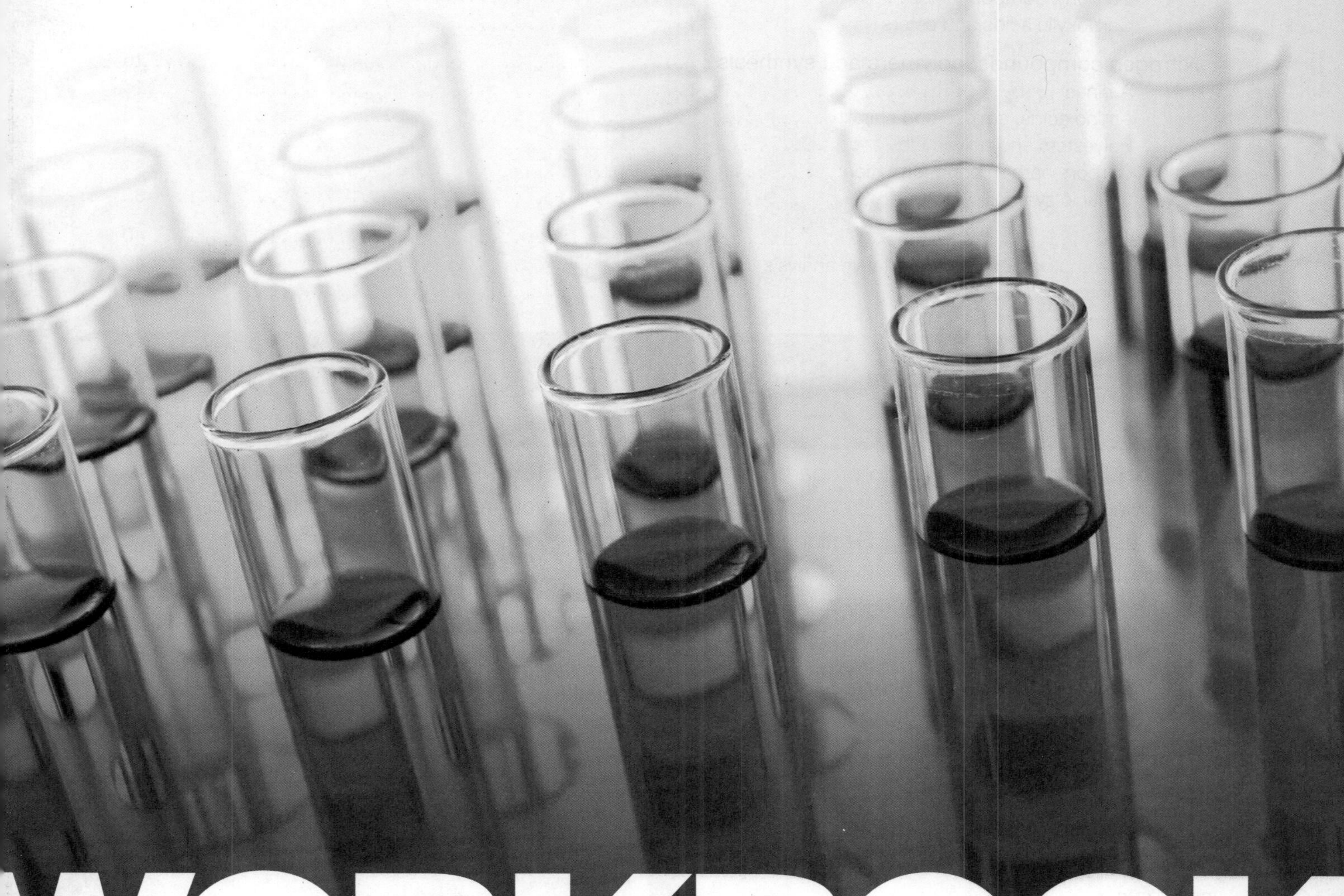

WORKBOOK

FOR THE 2015 SPECIFICATIONS

Chemistry A

Organic chemistry and analysis

John Older and Mike Smith

Contents

WORKBOOK

Module 6 Organic chemistry and analysis

1 **This workbook will** help you to prepare for the following exams:
- OCR Chemistry A-level Paper 1 Section B: the exam is 2 hours 15 minutes long, worth 100 marks overall and 37% of your A-level.
- OCR Chemistry A-level Paper 3: the exam is 1 hour 30 minutes long, worth 70 marks and 26% of your A-level.

2 **For each topic** there are:
- stimulus materials, including key terms and concepts
- questions testing key terms and concepts
- both short- and long-answer questions
- questions that test your mathematical skills
- questions that test your ability to handle A* questions
- space for you to write your answers.

3 **Answering the questions** will help you to build your skills and meet the assessment objectives AO1 (knowledge and understanding), AO2 (application) and AO3 (analysis, interpretation and evaluation).

4 **You still need to** read your textbook and refer to your revision guides and lesson notes.

5 **Marks available** are indicated for all questions so that you can gauge the level of detail required in your answers.

Timings are given for the exam-style questions to make your practice as realistic as possible.

Answers are available at:
www.hoddereducation.co.uk/workbookanswers

The content of this module assumes knowledge and understanding of the chemical concepts encountered in Module 2: Foundations in chemistry and Module 4: Core organic chemistry. It is anticipated that synoptic questions will be set that link together the chemistry in this module and in Modules 2 and 4.

3 a State the reagents and conditions required to convert methylbenzene into 1,4-dimethylbenzene. (AO2) 2 marks

Reagents ....

....

Conditions ....

....

Methylbenzene → 1,4-methylbenzene

b Describe, with the aid of curly arrows, the mechanism for this reaction. (AO2, AO3) 5 marks

Step 1:

....

Step 2:

....

Step 3:

....

4 Benzene, phenol and cyclohexene all react with bromine. Compare and contrast the reactions of bromine (AO1)

a with benzene and with phenol 3 marks

....

....

....

....

....

b **with benzene and with cyclohexene** **3 marks**

....

....

....

....

....

5 **a** **State *three* pieces of evidence that support the delocalised model for the bonding in benzene. (AO3)** **3 marks**

....

....

....

....

....

....

b **Suggest why the 2-, 4- and 6-directing effect of electron-releasing groups such as –OH and $–NH_2$ supports the Kekulé model. (AO3)** **4 marks**

....

....

....

....

....

....

Exam-style questions

34

1 **Phenol is manufactured using the cumene process. This involves two stages.** 14

a **The mechanism for stage 1 is outlined below.**

$$H_3C{-}CH{=}CH_2 + H^+ \longrightarrow H_3C{-}\overset{+}{C}H{-}CH_3$$

$$C_6H_6 + H_3C{-}\overset{+}{C}H{-}CH_3 \longrightarrow \text{Intermediate } [C_6H_6{-}CH(CH_3)_2]^+ \; + \ldots \longrightarrow \text{Cumene } C_6H_5CH(CH_3)_2 + \ldots\ldots$$

Intermediate   Cumene   Co-product

i **In the mechanism above, add curly arrows where appropriate and identify the co-product formed.** 4 marks

ii **Explain the *role* of the H^+ in stage 1.** 3 marks

....

....

....

....

iii **1-Phenylpropane, $C_6H_5CH_2CH_2CH_3$, is also formed as a minor product in stage 1. Suggest how.** 2 marks

....

....

....

....

b Stage 2 involves the atmospheric oxidation of cumene to form phenol and one other organic product.

i **The mass spectrum and infrared spectrum of the other organic product are shown below.**

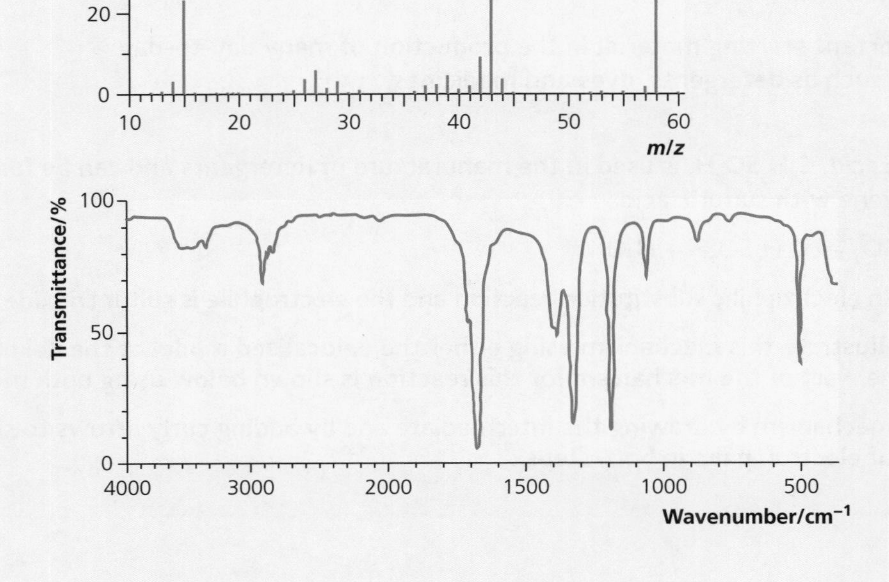

Use the spectra to identify the other organic product. **4 marks**

....

....

....

....

....

ii **Construct an equation for the atmospheric oxidation of cumene.** **1 mark**

....

c **Phenol can also be manufactured by reacting chlorobenzene with sodium hydroxide, as shown below.**

$$C_6H_5Cl + OH^- \xrightarrow[15\,000\ \text{kPa}]{400^\circ C} C_6H_5OH + Cl^-$$

i **What kind of reagent is OH^- in this reaction?** **1 mark**

....

ii **Suggest why the conditions used are so harsh.** **1 mark**

....

....

....

2 **Benzene is an important starting material in the production of many day-to-day essential products such as detergents, dyes and medicines.**

a **Benzenesulfonic acid, $C_6H_5SO_3H$, is used in the manufacture of detergents and can be formed by reacting benzene with sulfuric acid:**

$$C_6H_6 + H_2SO_4 \rightarrow C_6H_5SO_3H + H_2O$$

This reaction is an electrophilic substitution reaction and the electrophile is sulfur trioxide, SO_3.

It is possible to illustrate this mechanism using either the delocalised model or the Kekulé model of benzene. Part of the mechanism for this reaction is shown below using both models.

Complete each mechanism by drawing the intermediate and by adding curly arrows to show the movement of electron pairs in both steps.

i **delocalised model** **4 marks**

$O^{\delta-}$ $S^{\delta+}$ O O

Step 1 → **Intermediate** **Step 2** → O^- O=S=O + H^+

ii **Kekulé model** **4 marks**

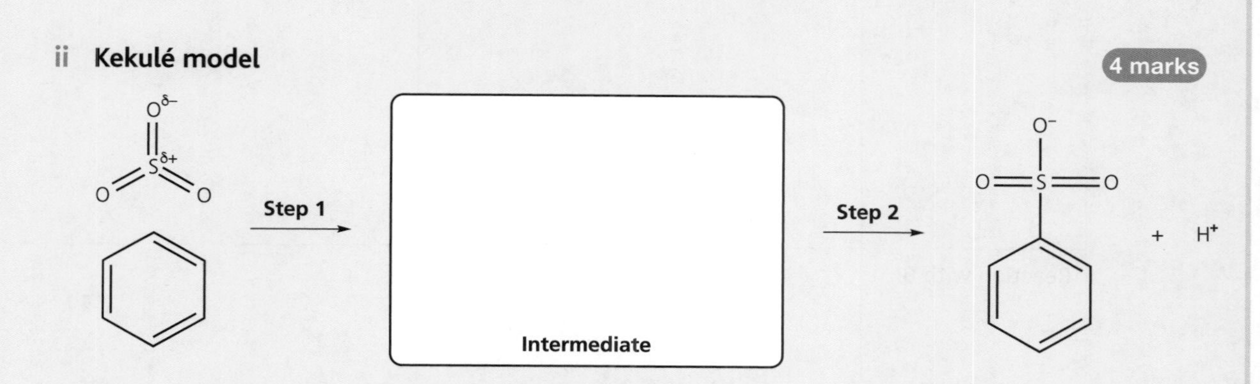

b **Aromatic compounds such as 4-chloronitrobenzene, $ClC_6H_4NO_2$, are intermediates in the manufacture of dyes. When chlorine is substituted onto the benzene ring, it is 2,4-directing. When the nitro group is substituted onto the benzene ring, it is 3-directing.**

Chemicals used in stage 1 ....

....

Chemicals used in stage 2 ....

....

i **State the chemicals used in each stage of the reaction and draw the structure of the organic intermediate.** **3 marks**

Explain the sequence of reactions. **2 marks**

....

....

....

....

ii **The organic product from stage 2 is then reduced to form $ClC_6H_4NH_2$. Write an equation for this reduction. Use [H] to represent the reducing agent.** **2 marks**

....

c **Paracetamol has the structure shown.**

Paracetamol

i **Separate samples of paracetamol are reacted with sodium and with bromine. Draw the structures of possible organic products and state what, if anything, you would expect to see.**

4 marks

	Organic product	**Observation (if any)**
Reaction with Na		
Reaction with Br_2		

ii **The infrared spectrum of paracetamol is shown below.**

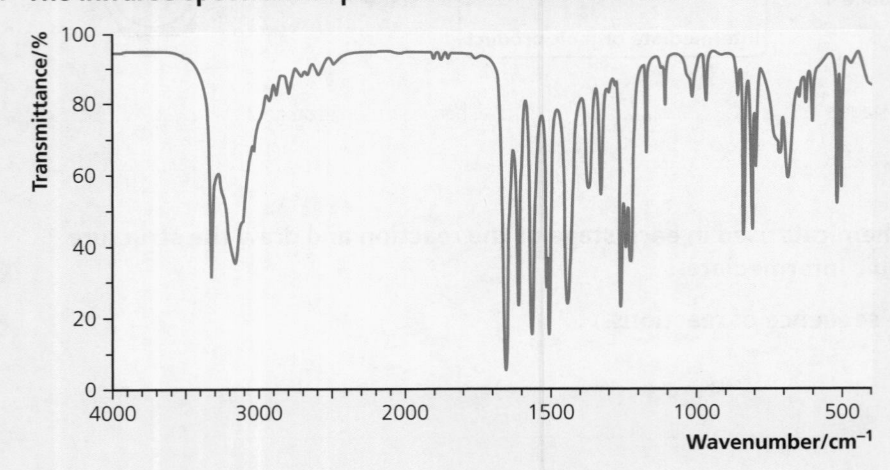

Label the peaks due to:

3 marks

- **the N–H bond**
- **the C=O bond**
- **the O–H bond**

Carbonyl compounds

You should, from Module 4, be aware of the oxidation of alcohols to form aldehydes, ketones and carboxylic acids. Aldehydes and ketones both contain the polar carbonyl group and therefore have some reactions in common. But aldehydes are readily oxidised and ketones are resistant to oxidation, so that aldehydes and ketones can be easily distinguished.

Reactions of carbonyl compounds

1 Complete the equation, name the mechanism and the organic product, and draw and explain the mechanism in the two reactions below. (AO2, AO3)

a — 7 marks

O + ... → OH CN

....

....

b — 7 marks

O + ... → OH

....

....

Characteristic tests for carbonyl compounds

2 a Describe how you would carry out the Tollens' test in the laboratory. (AO2, AO3) 3 marks

....

....

....

b Explain, with the aid of equations, how the Tollens' reagent would distinguish between propanal and propanone. (AO2, AO3) 3 marks

....

....

....

Carboxylic acids and esters

Properties of carboxylic acids

The reactions of acids with metals and bases were first covered in Module 2.

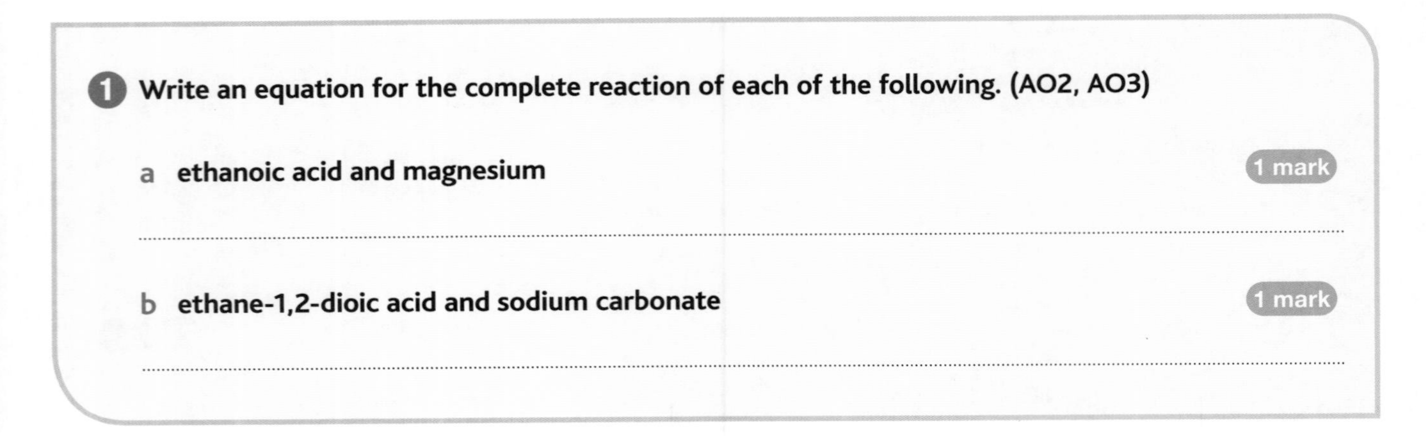

1 Write an equation for the complete reaction of each of the following. (AO2, AO3)

a ethanoic acid and magnesium 1 mark

....

b ethane-1,2-dioic acid and sodium carbonate 1 mark

....

Esters and acyl chlorides

You should be familiar with carboxylic acids and a series of acid derivatives, including esters, acid anhydrides, acyl chlorides and amides.

2 The scheme below shows the reactions of propanoic acid and some of its derivatives.

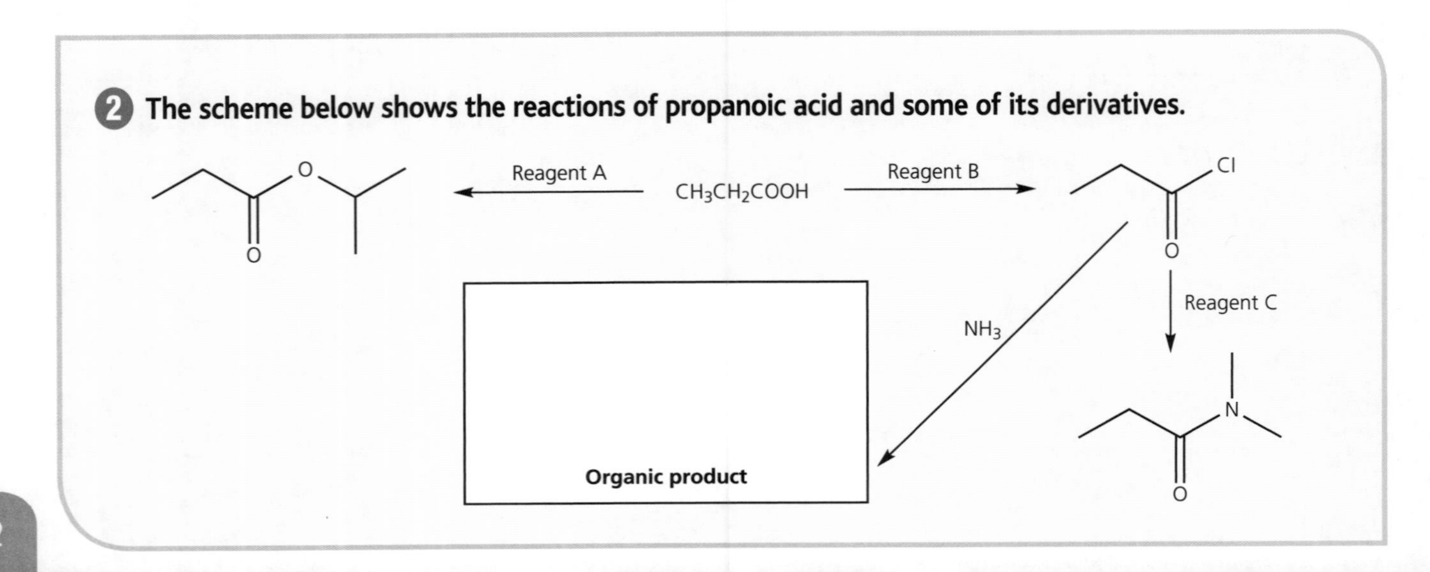

Give the name and structural formula of: (AO2, AO3)

a reagent A 2 marks

b reagent B 2 marks

c reagent C 2 marks

d What is the name of the organic product? Draw its displayed formula in the box in the scheme on p. 12. 2 marks

Exam-style questions

28

10

1 In an experiment, 6.90 g of ethanol was refluxed with an excess of ethanoic acid and concentrated sulfuric acid. The apparatus was then rearranged for distillation and the impure product was shaken with aqueous sodium carbonate. A vigorous effervescence was observed. The organic liquid was separated from the aqueous layer and redistilled. The fraction between 75°C and 79°C was collected. In total, 5.28 g of the ester ethyl ethanoate was collected.

Some relevant data for the compounds involved are shown in the table.

	Boiling point/°C	Density/g cm^{-3}
Ethanol	78	0.79
Ethanoic acid	118	1.05
Ethyl ethanoate	77	0.90

a i Write an equation for the reaction. 1 mark

ii What is the purpose of the sulfuric acid? 1 mark

iii Why was the sulfuric acid concentrated and not aqueous? 1 mark iv Calculate the minimum mass of ethanoic acid so that it is equimolar with the ethanol. **2 marks**

v Explain why it was necessary to reflux the mixture before rearranging the apparatus for distillation. **2 marks**

....

....

....

....

b i Write an ionic equation for the reaction that occurred when the impure product was shaken with aqueous sodium carbonate. **1 mark**

....

ii Suggest how the organic layer was separated from the aqueous layer. **1 mark**

....

....

c i Calculate the percentage yield. **2 marks**

ii Suggest *one* way that might improve the percentage yield. Justify your suggestion. **2 marks**

....

....

....

2 Ethanoic anhydride, $CH_3COOCOCH_3$, reacts with compound A to produce an ester and ethanoic acid. The atom economy for the ester is 69.39%. Identify the ester and compound A. Show all of your working. 8 **8 marks**

3 **4-Nitrophenol can be produced from 4-bromophenol. The electrophile is $^{+}NO_2$ and the bromine is substituted as shown below.**

OH, Br $\quad$ $^{+}NO_2$ ⟶ OH, +, Br, NO_2 ⟶ OH, NO_2 $\quad$ + ...

a **Complete the mechanism above by adding appropriate curly arrows and by identifying the missing product.** **3 marks**

b **The scheme below shows some reactions of 4-nitrophenol and 4-aminophenol.**

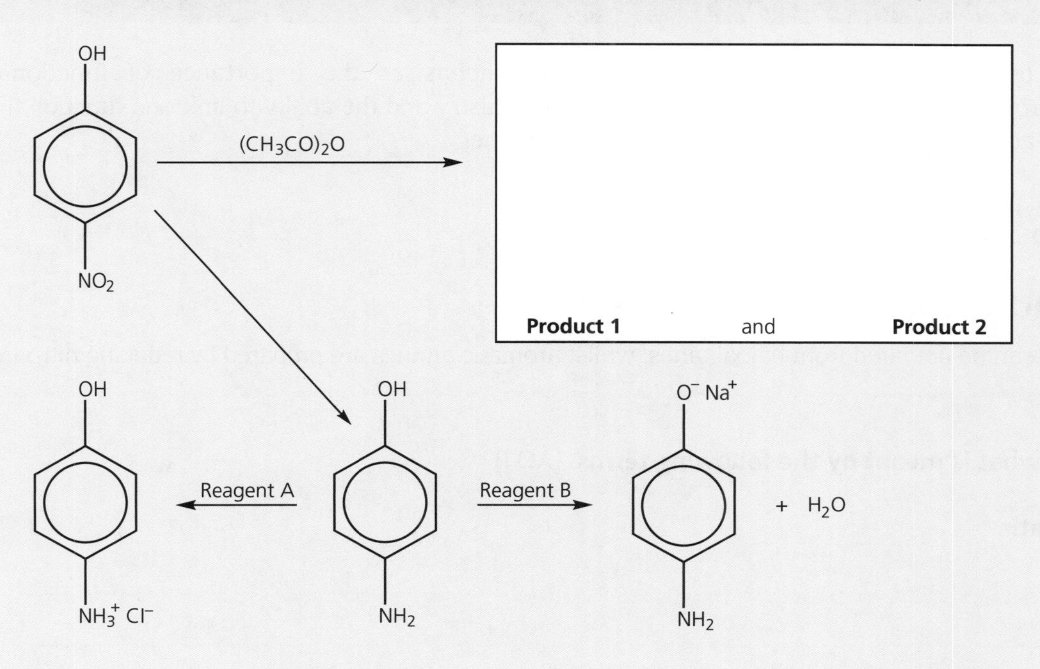

i **Identify the two reagents, A and B.** 2 marks

Reagent A is ....

Reagent B is ....

ii **4-Aminophenol is produced by the reduction of 4-nitrophenol. Write an equation for this reaction. Use [H] to represent the reducing agent.** 2 marks

....

iii **Draw products 1 and 2 in the box on p. 15.** 2 marks

c **4-Aminophenol reacts with bromine to produce a mixture of isomers. Draw all possible isomers with the molecular formula $C_6H_5NOBr_2$.** 4 marks

Nitrogen compounds, polymers and synthesis

This section focuses on amines, amides and amino acids. It also introduces condensation polymers, which are compared to addition polymers, studied in Module 4. It emphasises the importance of functional group chemistry and the ability to link one functional group to another.

Amines

Basicity and preparation of amines

Aliphatic amines can be prepared from haloalkanes, whilst aromatic amines are prepared by reducing nitroarenes.

1 Explain what is meant by the following terms. (AO1)

a **aliphatic** 1 mark

....

....

b aromatic 

....

....

2 **Ethylamine, $CH_3CH_2NH_2$, can be prepared by reacting chloroethane with *excess* ammonia in a suitable solvent.**

a i State a suitable solvent. (AO1) 

....

ii Why is water not a suitable solvent? (AO2) 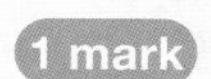

....

....

b i The mechanism for the reaction is outlined below.

CH_3CH_2Cl + NH_3 → $CH_3CH_2\overset{+}{N}H_3$ Cl^- → $CH_3CH_2NH_2$ + H^+ + Cl^-

On the mechanism above, add curly arrows and show relevant dipoles and lone pairs of electrons. (AO3). 

ii Complete the equation below for this reaction. (AO2) 

CH_3CH_2Cl + ...NH_3 → $CH_3CH_2NH_2$ + ...

iii If excess ammonia is not used, compound X, molecular formula $C_4H_{11}N$, is also formed. Draw the structural formula of compound X. (AO2, AO3) 

3 **The following aromatic compounds were each reduced using tin and concentrated hydrochloric acid. Draw the product in the box and balance each equation. (AO2)** 

a

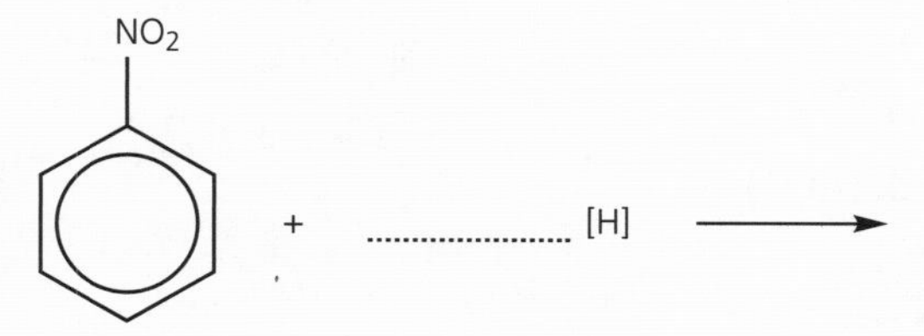

\+ ....

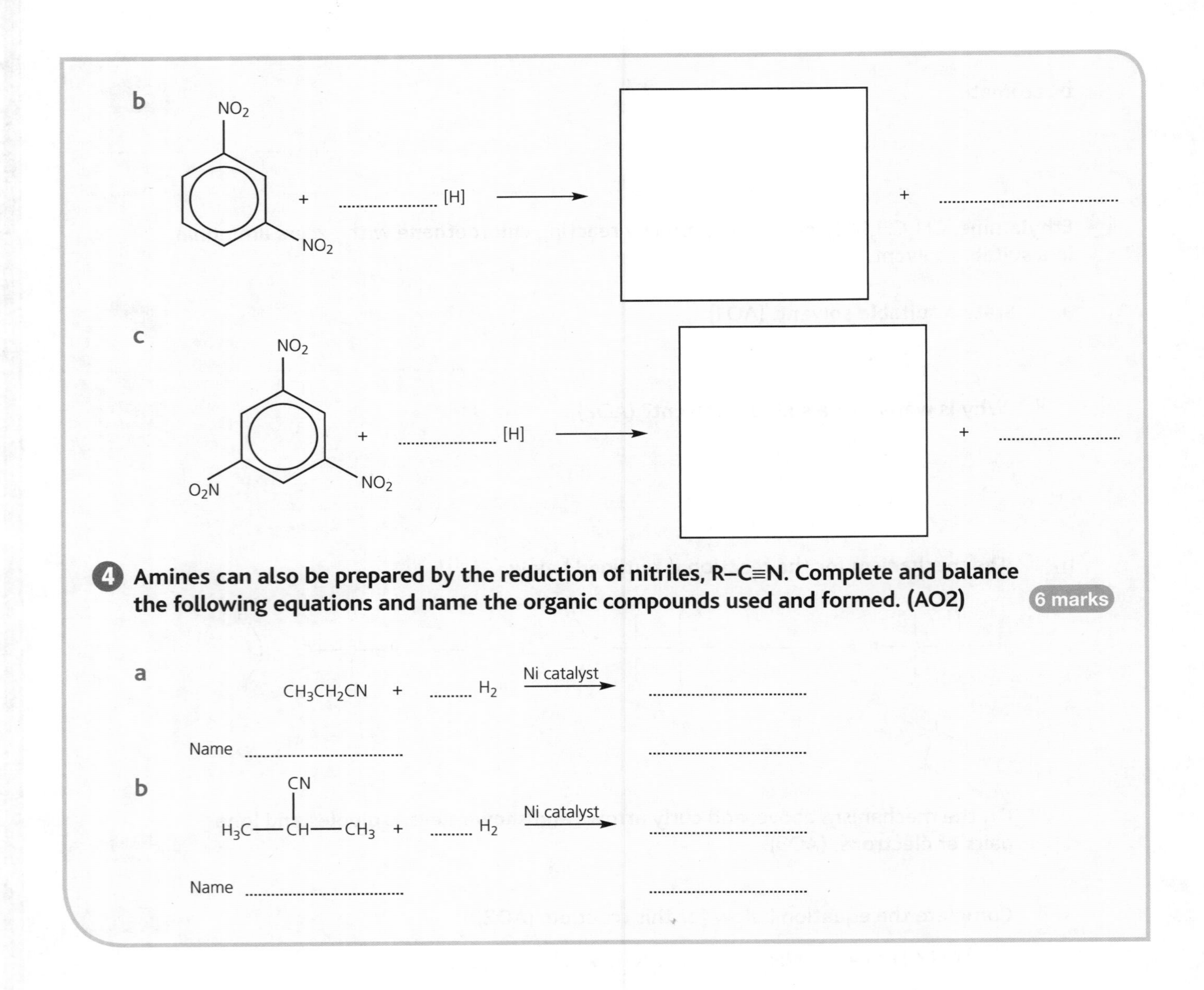

4 **Amines can also be prepared by the reduction of nitriles, R–C≡N. Complete and balance the following equations and name the organic compounds used and formed. (AO2)** 6 marks

a CH_3CH_2CN + ... H_2 $\xrightarrow{\text{Ni catalyst}}$ ....

Name ....

b $H_3C—CH(CN)—CH_3$ + ... H_2 $\xrightarrow{\text{Ni catalyst}}$ ....

Name ....

Amino acids, amides and chirality

Reactions of amino acids

Amino acids are naturally occurring compounds that contain two functional groups and display properties of both acids and bases.

1 a **What is meant by the term 'α-amino acid'? (AO1)** 1 mark

....

....

b **State the general formula of an α-amino acid. (AO1)** 1 mark

....

Amides and chirality

2 **Explain what is meant by the term 'chiral'. (AO1)** 1 mark

....

....

3 a **The table below shows the R group in three different amino acids. Complete the table by drawing three-dimensional structures of both optical isomers. (AO2)**

Use wedge-shaped bonds such as

and

to show the three-dimensional arrangement around the central carbon atom.

	Amino acid	R group	Optical isomer 1	Optical isomer 2
i	Valine (val)	$-CH(CH_3)_2$		
ii	Lysine (lys)	$-(CH_2)_4NH_2$		
iii	Aspartic acid (asp)	$-CH_2COOH$		

b **Draw the organic product obtained when: (AO2, AO3)**

i **valine reacts with ethanol in the presence of concentrated sulfuric acid**

ii **lysine reacts with hydrochloric acid**

iii aspartic acid reacts with ammonia. 1 mark

c The R group in glycine (gly) is H. Explain why glycine does not have optical isomers. (AO2) 1 mark

....

....

d Glycine (gly), molecular formula $C_2H_5NO_2$, reacts with valine, molecular formula $C_5H_{11}NO_2$, to produce a mixture of two structural isomers each with molecular formula $C_7H_{14}N_2O_3$.

i Construct an equation for this reaction using the molecular formulae. (AO3) 1 mark

....

ii Draw the two possible isomers that each have the molecular formula $C_7H_{14}N_2O_3$. (AO2, AO3) 2 marks

e The structure below shows a short section of a polypeptide made from the four amino acids used in this question.

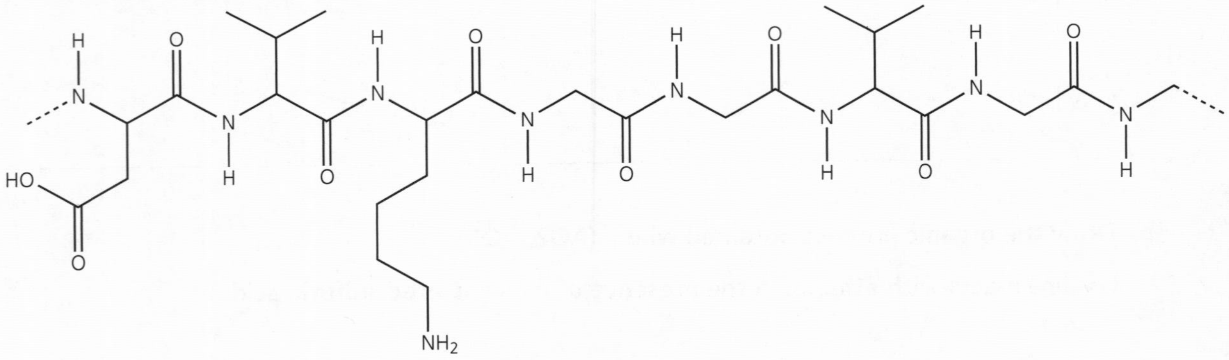

Using the three-letter abbreviation for each amino acid, write the sequence of the amino acids in the section shown. (AO2) 1 mark

....

4 Identify the structural feature that gives rise to the stereoisomerism in each of the following. Circle any double bonds that lead to *E*/*Z* isomerism. Put a star * next to any chiral centres. Delete the incorrect options. (AO2/AO3) 8 marks

a  *E*/*Z* or optical or both or neither (*delete as appropriate*)

b 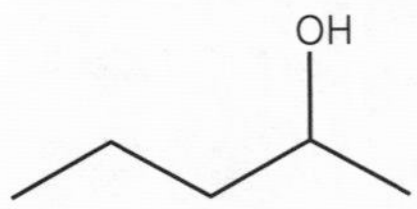 *E*/*Z* or optical or both or neither (*delete as appropriate*)

c

E/Z or optical or both or neither (*delete as appropriate*)

d

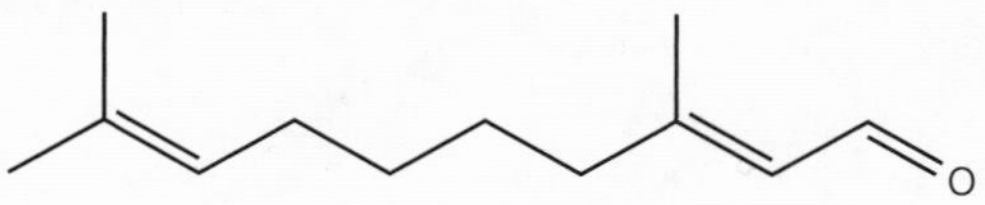

O

E/Z or optical or both or neither (*delete as appropriate*)

e

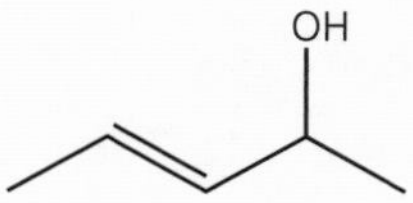

OH

E/Z or optical or both or neither (*delete as appropriate*)

f

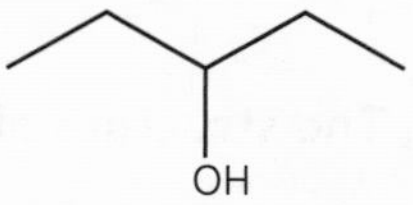

OH

E/Z or optical or both or neither (*delete as appropriate*)

g

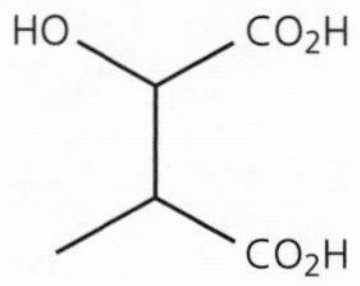

HO CO_2H

CO_2H

E/Z or optical or both or neither (*delete as appropriate*)

h

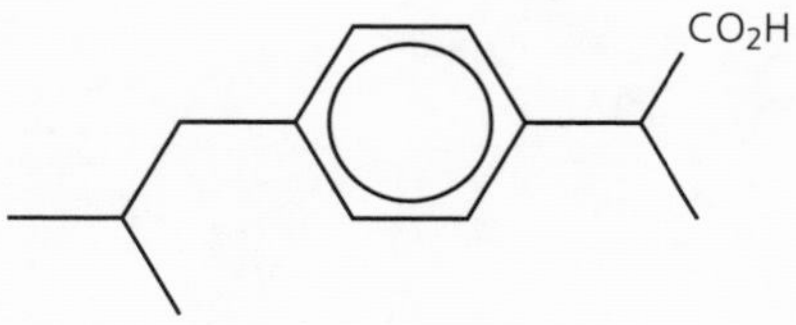

CO_2H

E/Z or optical or both or neither (*delete as appropriate*)

5 **Draw each of the following. (AO2)** **4 marks**

a **methanamide**

b **ethylamine**

c **phenylamine**

d **propanamide**

Exam-style questions

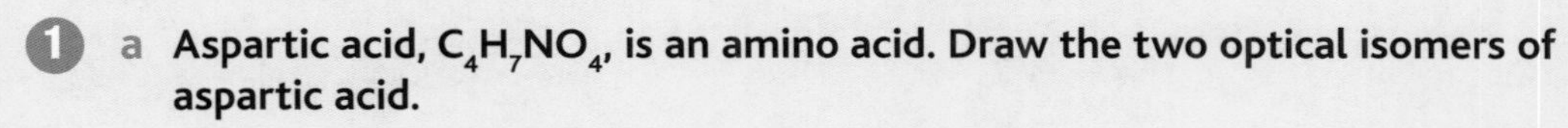

1 a **Aspartic acid, $C_4H_7NO_4$, is an amino acid. Draw the two optical isomers of aspartic acid.** **2 marks**

b **Aspartic acid reacts with excess NaOH(aq) to form a salt. Draw the structural formula of the salt.** 2 marks

c **The artificial sweetener, aspartame, can be prepared from aspartic acid. The structure of aspartame is shown below.**

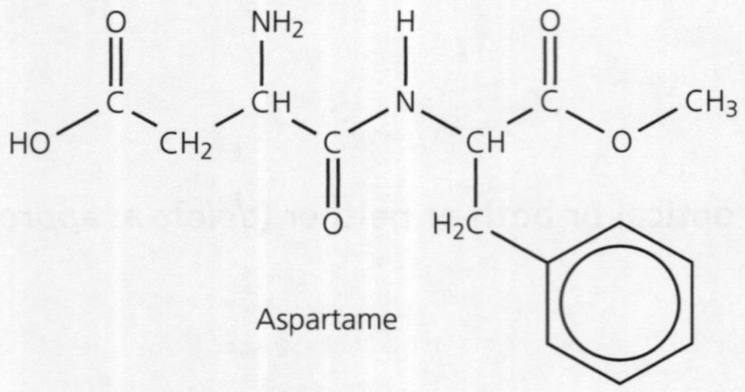

Aspartame

i **Draw the structure of a compound that could react with aspartic acid to form aspartame.** 2 marks

ii **List the functional groups in aspartame.** 2 marks

....

....

d **Draw the organic product or products formed when:**

i **aspartic acid reacts with excess $SOCl_2$** 2 marks

ii **aspartame is refluxed with excess dilute hydrochloric acid**

2 a **Tartronic acid is a naturally occurring acid and contains, by mass, 30.00% C, 3.33% H and 66.7%. The molar mass of tartronic acid is 120 g mol^{-1}. Calculate the molecular formula of tartronic acid.**

3 marks

b **Malic acid, shown below, has a similar structure to tartronic acid.**

i **What is the molecular formula of malic acid?**

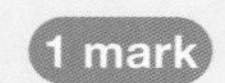

ii **Deduce the skeletal structure of tartronic acid.**

c **Malic acid reacts with aqueous sodium carbonate solution and also reacts with hot acidified potassium dichromate solution. Draw the skeletal formula of the following:**

i **the organic salt formed when malic acid reacts with aqueous sodium carbonate** 1 mark

ii **organic product formed when malic acid reacts with hot acidified potassium dichromate** 1 mark

d **The scheme below shows some reactions of malic acid.**

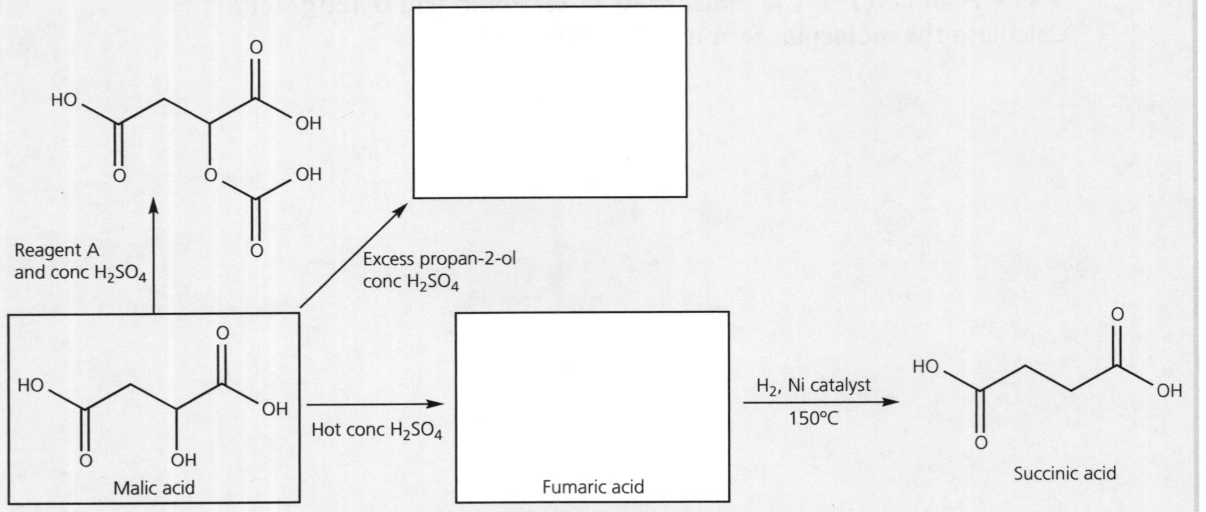

Reagent A and conc H_2SO_4

Excess propan-2-ol conc H_2SO_4

Malic acid

Hot conc H_2SO_4

Fumaric acid

H_2, Ni catalyst 150°C

Succinic acid

i **Identify reagent A.** 1 mark

....

ii **Write an equation for the reaction between malic acid and excess propan-2-ol. Show clearly the structure of the organic product.** 3 marks iii **Deduce the structural formula of fumaric acid.** 

e **Malic acid and fumaric acid each has a stereoisomer.**

i **State the type of stereoisomerism in malic acid and identify the key structural feature that gives rise to the stereoisomerism. Draw malic acid and its stereoisomer.** 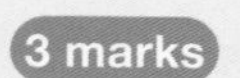

....

....

ii **State the type of stereoisomerism in fumaric acid and identify the key structural feature that gives rise to the stereoisomerism. Draw fumaric acid and its stereoisomer.** 

....

....

3 **Ethanoic anhydride can be used in the laboratory synthesis of both aspirin and paracetamol.** 

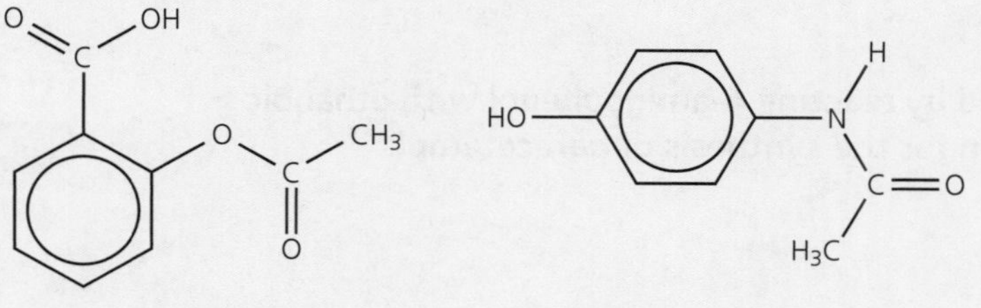

Aspirin

Paracetamol

a **Draw ethanoic anhydride.** **1 mark**

b **In a synthesis, 1.05 g of 2-hydroxybenzoic acid was mixed with 2 cm³ of ethanoic anhydride and warmed gently until all the solid had reacted. The mixture was then allowed to cool in an ice bath until precipitation was complete. The product was filtered using a Büchner funnel, washed with cold water and allowed to dry overnight. A total of 0.98 g of aspirin was produced.**

2-hydroxybenzoic acid —Ethanoic anhydride→ Aspirin + Organic co-product

i **Identify the organic co-product.** **1 mark**

....

ii **Draw a labelled sketch to show how the product was filtered using a Büchner funnel.** **2 marks**

iii **Calculate the percentage yield.** **5 marks**

c **Paracetamol can also be prepared by reacting 4-aminophenol with ethanoic anhydride. Construct an equation for the synthesis of paracetamol.** **2 marks**

d **The purity of the samples of aspirin and paracetamol can be assessed by measuring their melting points. If the samples are impure, explain how this would affect the melting points.** 2 marks

....

....

....

....

e **Aspirin and paracetamol both react with NaOH. Draw the products formed when NaOH is refluxed with:**

i **aspirin** 3 marks

ii **paracetamol** 2 marks

f **Identify a reagent that would react with aspirin but not with paracetamol. Explain your answer.** 2 marks

....

....

....

g **Identify a reagent that would react with paracetamol but not with aspirin. Explain your answer.** 2 marks

....

....

....

Polyesters and polyamides

Condensation polymers

Addition polymerisation was covered in Module 4. You should be able to deduce the structure of a polymer from a known monomer as well as being able to work out the monomer that could have been used to make a known polymer.

Condensation polymers can be made from two different monomers or from a single monomer that contains two different functional groups.

Polyesters and polyamides can both be hydrolysed using either acid or base catalysts.

1 **Draw the monomer and two repeat units of the polymer that could be made from: (AO2)** 2 marks

a **but-2-ene**

b **phenylethene**

2 **Deduce the monomer that was used to make the polymer below. (AO2)** 1 mark

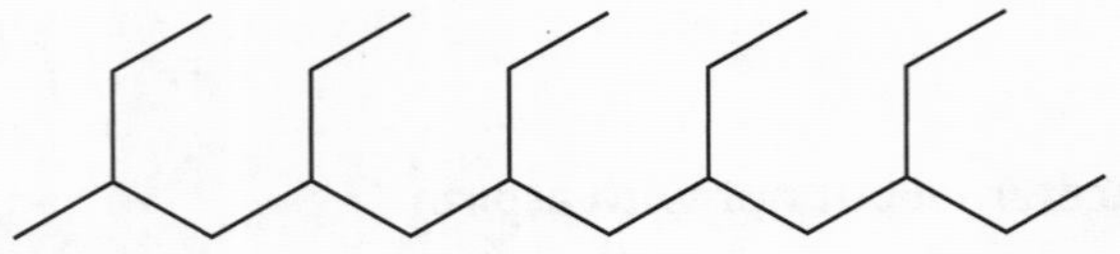

....

3 Polyesters and polyamides are condensation polymers. Explain what is meant by the term 'condensation polymer'. (AO1) **2 marks**

....

....

....

....

....

4 a Draw each of the following: (AO1, AO2) **4 marks**

i ethane-1,2-diol

ii benzene-1,4-dioyl chloride

iii 1,4-diaminohexane

iv hexanedioic acid

b Draw each of the following: (AO1, AO2) **4 marks**

i 2-hydroxypropanoic acid

ii 3-hydroxypentanoic acid

iii aminoethanoic acid (glycine)

iv 2-amino-3-methylbutanoic acid (valine)

5 Short sections of two polyesters are shown below. For each polymer, identify the repeat unit by drawing a circle (or oval) around it, and then identify and draw the two monomers. (AO2, AO3)

a — 3 marks

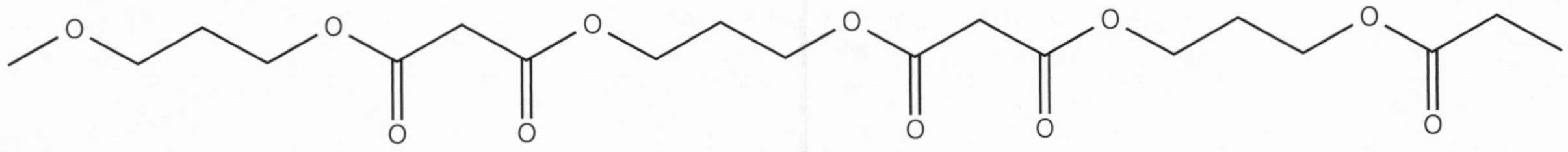

Monomer 1	Monomer 2

b — 3 marks

Monomer 1	Monomer 2

6 Short sections of two polymers are shown below.

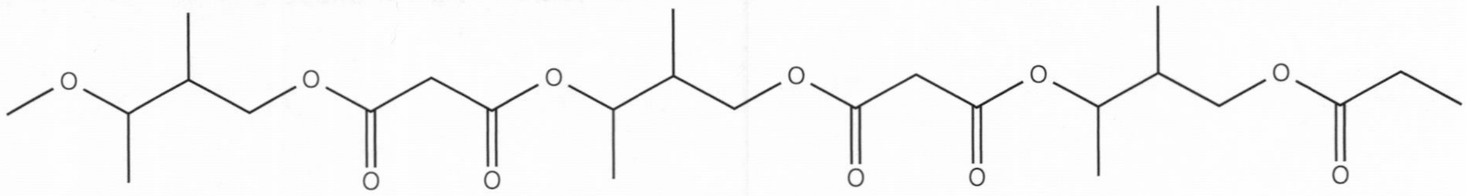

Polymer A

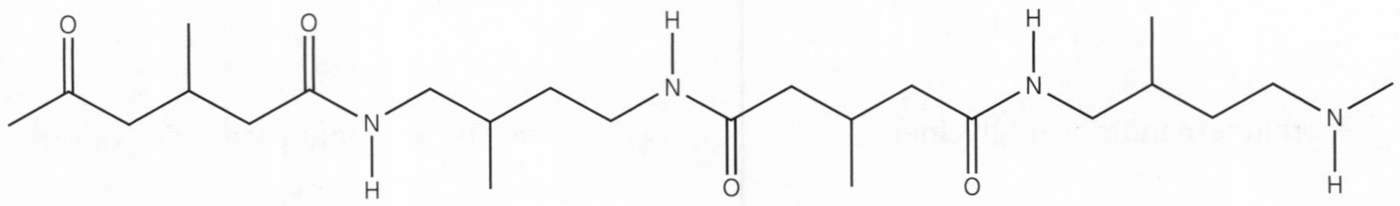

Polymer B

Use skeletal formulae to identify the organic products formed when each polymer is hydrolysed in acidic conditions and also in alkaline conditions. (AO2, AO3)

a acid hydrolysis of polymer A **2 marks**

b alkaline hydrolysis of polymer A **2 marks**

c acid hydrolysis of polymer B **2 marks**

d alkaline hydrolysis of polymer B **2 marks**

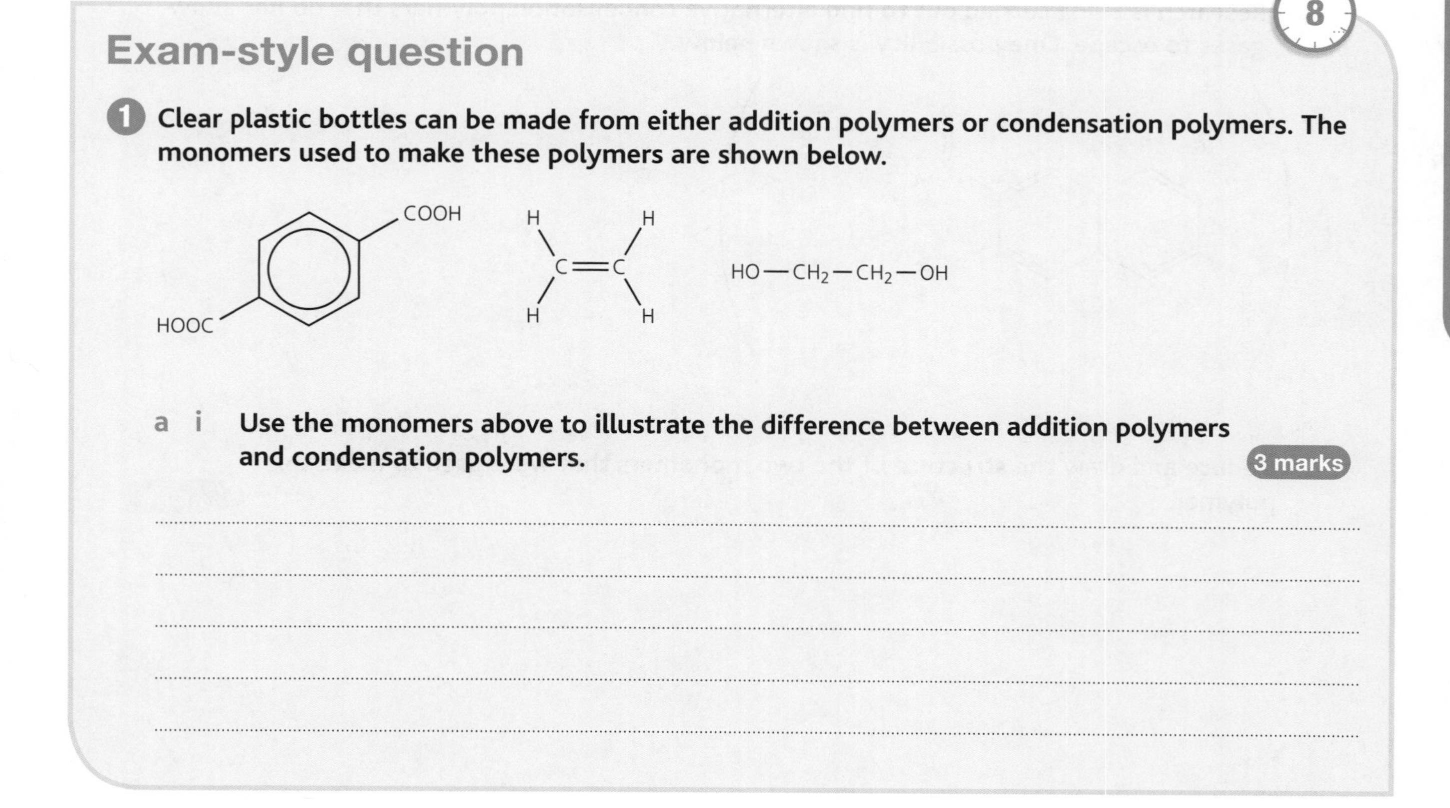

Exam-style question

1 **Clear plastic bottles can be made from either addition polymers or condensation polymers. The monomers used to make these polymers are shown below.**

a i **Use the monomers above to illustrate the difference between addition polymers and condensation polymers.** **3 marks**

....

....

....

....

....

ii **Draw the repeat unit in the condensation polymer.** 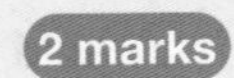

b **Bottles made from the condensation polymer above often contain fizzy drinks. However, the polymer allows gas to slowly escape and the fizzy drink goes flat. One solution to this is to add a layer of another polymer, polyvinyl alcohol (PVOH), to reduce the gas leakage. PVOH is manufactured in two stages:**

Stage 1 → addition polymerisation → ( ) → Stage 2 → repeat unit of PVOH

Repeat unit of intermediate polymer

Repeat unit of PVOH

i **In the space in the scheme above, draw the repeat unit of the intermediate polymer formed in stage 1.** 1 mark

ii **Stage 2 involves hydrolysis of the side chain. State the reagent and conditions required for stage 2.** 2 marks

....

....

c **Research is being carried out to find alternative condensation polymers that do not allow gases to escape. One possibility is shown below.**

CH_2 CH_2 n

Deduce and draw the structure of the two monomers that were used to make this polymer. 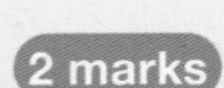

Carbon–carbon bond formation

Extending carbon chain length

When synthesising new compounds, it may be necessary to extend the carbon chain. This usually involves the formation of a nitrile. Nitriles can be formed by reacting cyanide ions, CN^-, with either haloalkanes or carbonyl compounds.

Carbon atoms can also be introduced by Friedel–Crafts reactions with aromatic compounds.

The nitrile group, R–C≡N, can be reduced and can be hydrolysed.

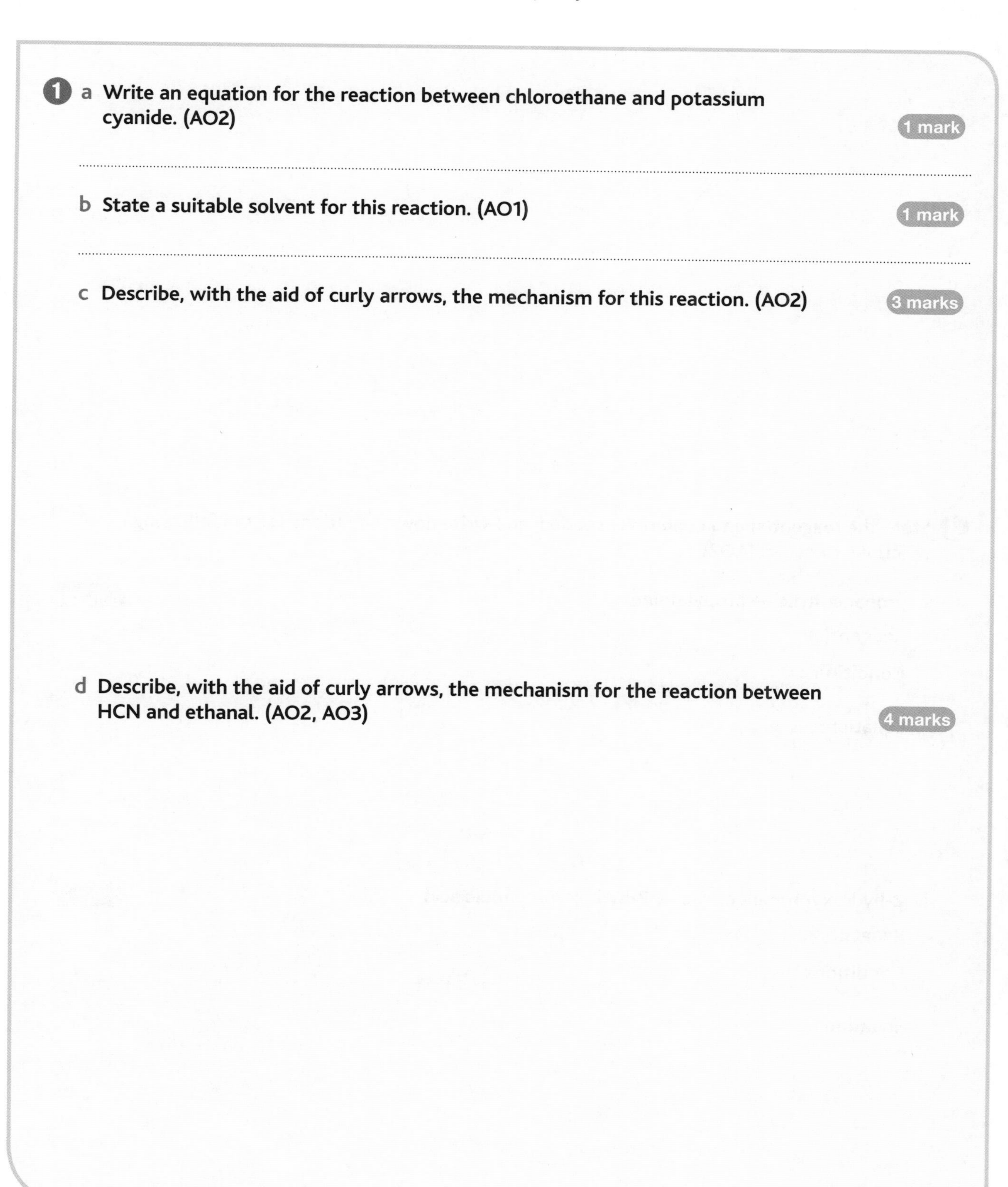

1 a **Write an equation for the reaction between chloroethane and potassium cyanide. (AO2)** 1 mark

....

b **State a suitable solvent for this reaction. (AO1)** 1 mark

....

c **Describe, with the aid of curly arrows, the mechanism for this reaction. (AO2)** 3 marks

d **Describe, with the aid of curly arrows, the mechanism for the reaction between HCN and ethanal. (AO2, AO3)** 4 marks

2 Benzene can be converted into phenylethanone, $C_6H_5COCH_3$. State the reagents and conditions needed for this reaction, and then write down the mechanism. (AO2) **6 marks**

Reagents: ....

Conditions: ....

Mechanism:

3 State the reagent(s) and conditions needed, and write down equations for the following reactions to occur: (AO2)

a propanenitrile → propylamine **3 marks**

Reagent(s): ....

Conditions: ....

Equation:

b 2-hydroxypropanenitrile → 2-hydroxypropanoic acid **3 marks**

Reagent(s): ....

Conditions: ....

Equation:

Organic synthesis

Practical skills

You should be aware of standard organic techniques and be familiar with the Quickfit apparatus necessary for reflux and for distillation.

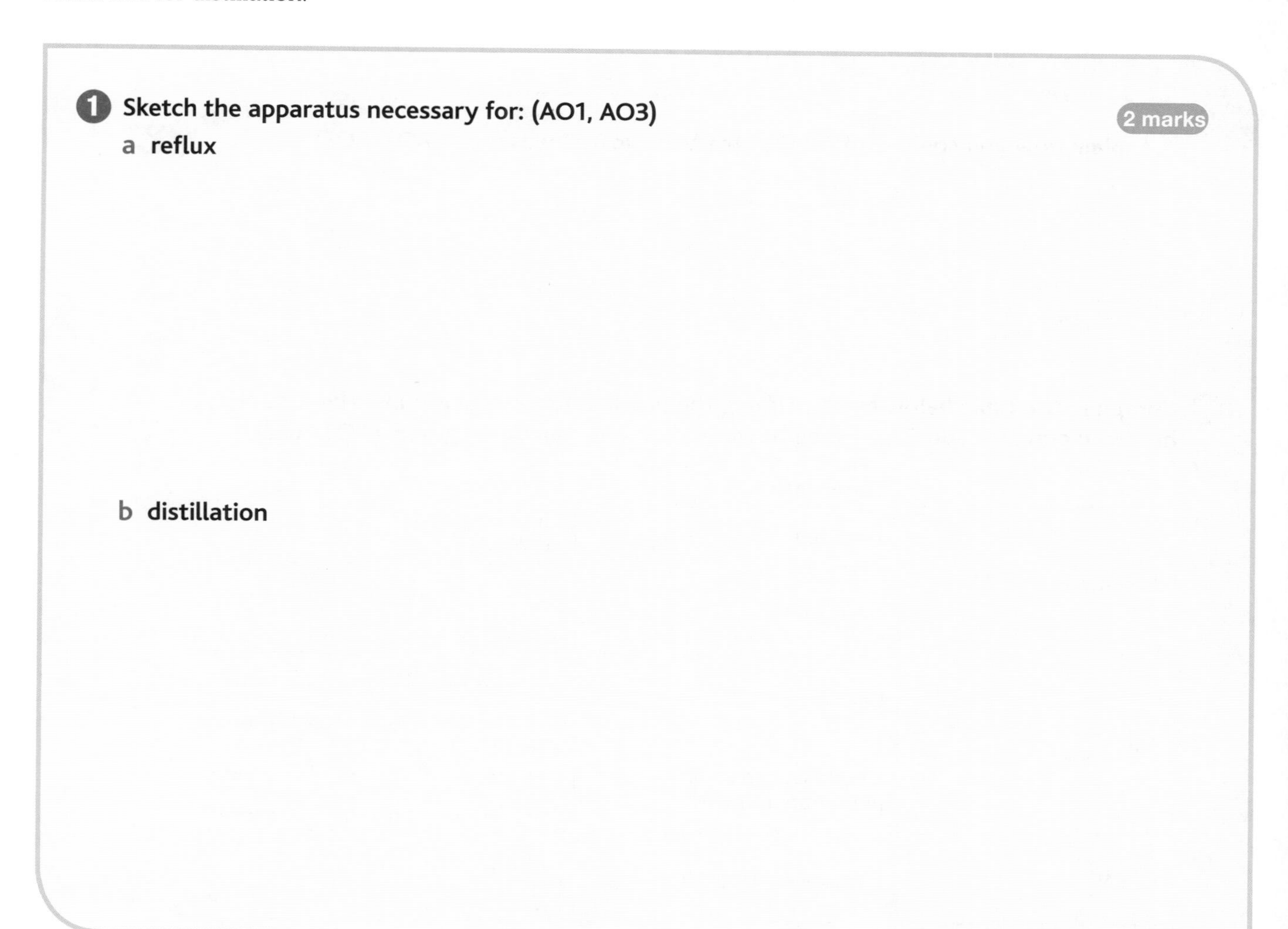

1 Sketch the apparatus necessary for: (AO1, AO3) 2 marks

a reflux

b distillation

Synthetic routes

In any organic synthesis, the product is unlikely to be pure. You should be aware of the techniques used for the purification of both liquid and solid products.

You are expected to be able to identify functional groups encountered in this module and in Module 4, and to be able to devise multi-stage synthetic routes linking together a variety of functional groups.

2 A sample of benzoic acid is contaminated with copper sulfate and with graphite.

a Use the information in the table below to describe how you could obtain a pure sample of the benzoic acid. (AO1, AO3) 4 marks

	Soluble in cold water?	Soluble in hot water?
Benzoic acid	No	Yes
Copper sulfate	Yes	Yes
Graphite	No	No

....

....

....

....

....

....

b Explain how you could confirm that the benzoic acid was pure. (AO1, AO3) **2 marks**

....

....

....

....

3 Complete the table below by identifying the possible functional groups. The first one has been done for you. There may be more than one functional group. (AO1, AO2) **10 marks**

Test	Observation	Inference
Add litmus	Turns red	Carboxylic acid or phenol
	Turns blue	....
Add Br_2	Decolorises	....
	Decolorises and white precipitate formed	....
Add Na	Effervescence/bubbles	....
Add $Na_2CO_3(aq)$	Effervescence/gas (CO_2) given off, bubbles, fizzes	....
In water bath at about 60°C with $AgNO_3(aq)$/ethanol	White, cream or yellow precipitate formed	....
Tollens' reagent/ $Ag^+(NH_3)_2$	Silver mirror formed	....
Heat with $H^+/Cr_2O_7^{2-}$	Colour changes from orange to green	....
Add water	White fumes of HCl given off	....
Warm with NaOH(aq)	Smell of NH_3 gas (which turns litmus blue)	....

4 Devise a two-stage synthesis to convert propanal into 1-bromopropane. For each stage, write an equation and state any conditions required. (AO2) 6 marks

Stage 1

....

....

....

Stage 2

....

....

....

5 Devise a two-stage synthesis to convert butan-1-ol into butan-2-ol. For each stage, write an equation and state any conditions required. (AO2) 6 marks

Stage 1

....

....

....

....

Stage 2

....

....

....

....

6 Devise a multi-stage synthesis to convert propene into 2-hydroxy-2-methylpropanoic acid, $CH_3C(CH_3)(OH)COOH$. For each stage, write an equation and state any conditions required. (AO2) 12 marks

Stage 1

....

....

....

....

Stage 2

....

....

....

....

Stage 3

....

....

....

....

Stage 4

....

....

....

....

10

Exam-style question

1 **Acetanilide can be prepared from benzene via a three-stage synthesis, via compounds A and B, as shown below.**

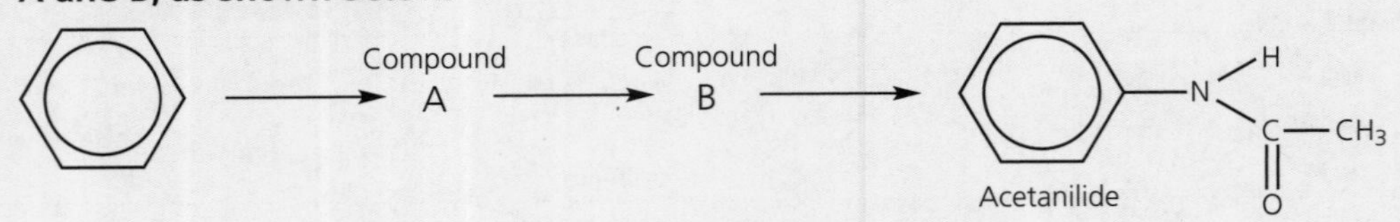

a **Compound B contains by mass 77.42% C, 15.05% N and 7.53% H. The infrared and mass spectra of compound B are shown below.**

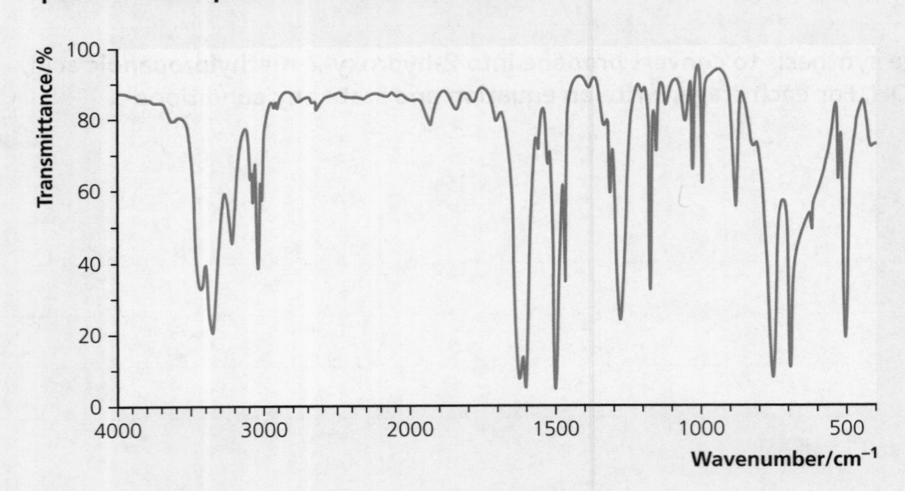

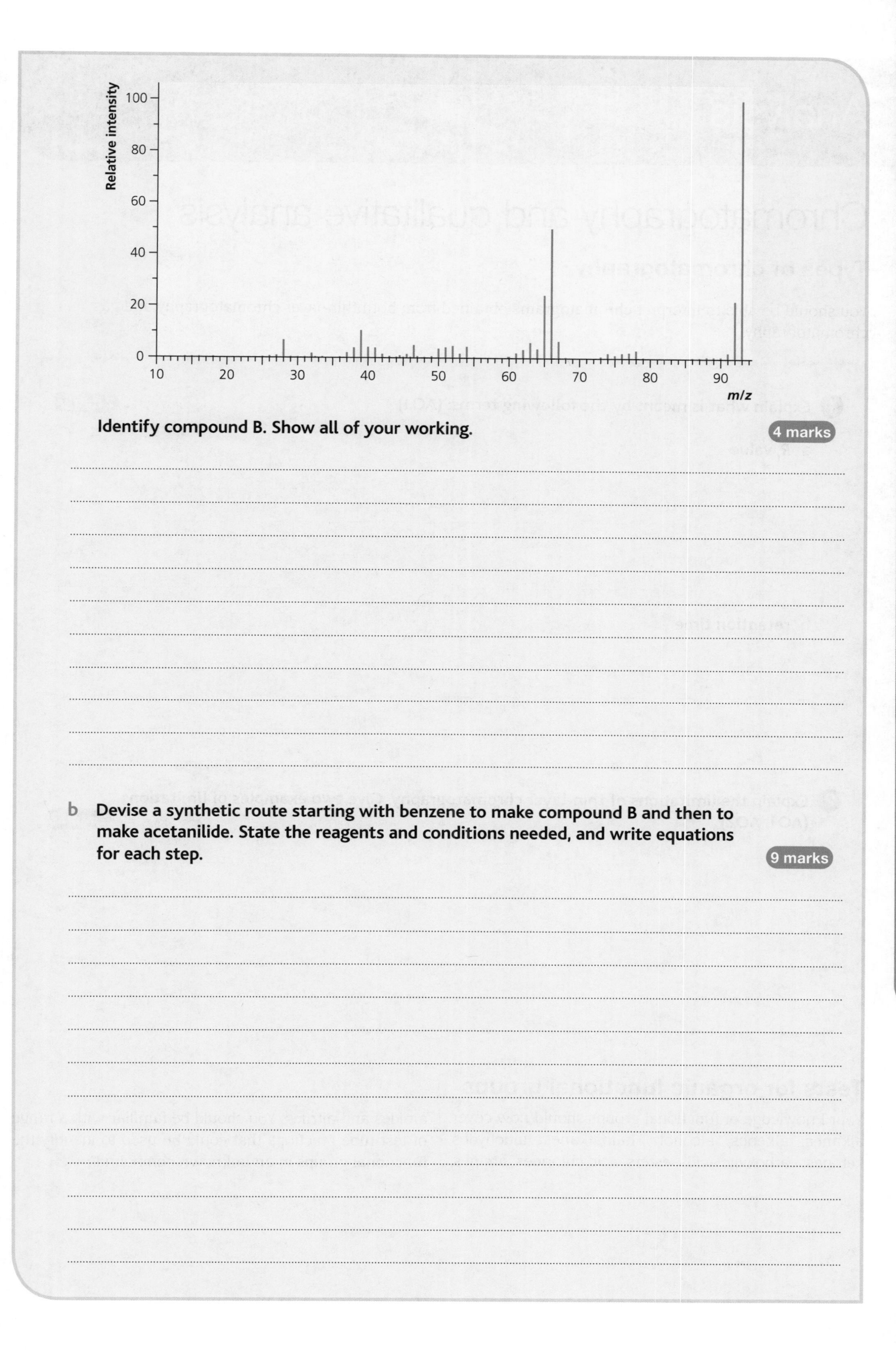

Identify compound B. Show all of your working. **4 marks**

b **Devise a synthetic route starting with benzene to make compound B and then to make acetanilide. State the reagents and conditions needed, and write equations for each step.** **9 marks**

Chromatography and qualitative analysis

Types of chromatography

You should be able to interpret chromatograms obtained from both thin-layer chromatography and gas chromatography.

1 Explain what is meant by the following terms: (AO1) 2 marks

a R_f value

....

....

....

....

b retention time

....

....

....

....

2 Explain the limitations of thin-layer chromatography. Give two examples of limitations. (AO1, AO2) 2 marks

....

....

....

....

....

Tests for organic functional groups

Your knowledge of functional groups should now cover: alkanes, alkenes, alcohols, haloalkanes, aldehydes, ketones, carboxylic acids, esters, acid chlorides, amines, amides and nitriles. You should be familiar with a range of test tube reactions that could be used to identify the functional groups in an unknown compound.

3 Complete the table below to provide chemical tests to identify each of the following functional groups. (AO1, AO2)

20 marks

Functional group		Reagent/conditions	Observation
Name	Group		
Alkene	C=C		
Haloalkane	R—Cl		
*Aldehyde	—C(=O)H		
*Ketone	—C(=O)—		
*Phenol	C_6H_5—OH		

Carboxylic acid	—C(=O)OH	....	....
Acid chloride	—C(=O)Cl	....	....

*Each of these functional groups requires a combination of two tests to confirm its presence.

Spectroscopy

NMR spectroscopy

In the assessment of ^{13}C NMR spectra, all different ^{13}C environments are considered as low-resolution and will appear as single peaks—there will be no splitting.

1 How many different ^{1}H environments and ^{13}C environments are there in the following. (AO2)

a **2 marks**

```
    H   H   H   H
    |   |   |   |
H — C — C — C — C — H
    |   |   |   |
    H   H   H   H
```

number of peaks in ^{1}H NMR = ...
number of peaks in ^{13}C NMR = ...

b **2 marks**

```
    H   H   H
    |   |   |
H — C — C — C — H
    |   |   |
    H   C   H
      / | \
     H  H  H
```

number of peaks in ^{1}H NMR = ...
number of peaks in ^{13}C NMR = ...

c **2 marks**

```
    H   H   O   H
    |   |   ‖   |
H — C — C — C — C — H
    |   |       |
    H   H       H
```

number of peaks in ^{1}H NMR = ...
number of peaks in ^{13}C NMR = ...

d **2 marks**

number of peaks in ^{1}H NMR = ...
number of peaks in ^{13}C NMR = ...

e **2 marks**

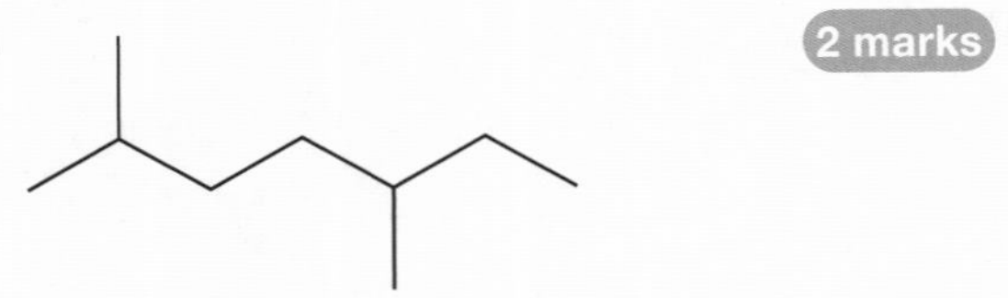

number of peaks in ^{1}H NMR = ...
number of peaks in ^{13}C NMR = ...

f **2 marks**

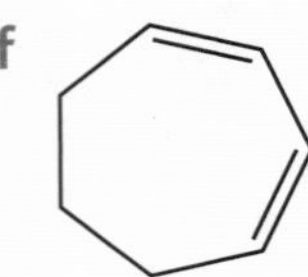

number of peaks in ^{1}H NMR = ...
number of peaks in ^{13}C NMR = ...

2 Each ^{13}C or ^{1}H environment is detected at a different frequency identified by its chemical shift, δ, which is compared to the chemical shift of the standard, tetramethylsilane, or TMS, $Si(CH_3)_4$.

a How many different hydrogen environments and carbon environments are present in $Si(CH_3)_4$? (AO2) **2 marks**

hydrogen environments = ...

carbon environments = ...

b Give *three* reasons why TMS is a good standard. (AO1, AO3) **3 marks**

....

....

....

....

....

....

c Use your data sheet to estimate the chemical shift, δ, of the highlighted (bold and underlined) $^{1}\underline{H}$ or $^{13}\underline{C}$ in each of the following. (AO1, AO2) **28 marks**

Identify the range for the $^{1}\underline{H}$ environment		Identify the range for the $^{13}\underline{C}$ environment	
Chemical environment	δ	**Chemical environment**	δ
$C\underline{\mathbf{H}}_3CH_2OH$		$\underline{\mathbf{C}}H_3CH_2OH$	
$C\underline{\mathbf{H}}_3COOH$		$CH_3\underline{\mathbf{C}}OOH$	
$C\underline{\mathbf{H}}_3CHO$		$CH_3\underline{\mathbf{C}}HO$	
$C_6H_5C\underline{\mathbf{H}}_3$		$\underline{\mathbf{C}}_6H_5CH_3$	
$CH_3CH_2O\underline{\mathbf{H}}$		$CH_3\underline{\mathbf{C}}H_2NH_2$	
$CH_3C\underline{\mathbf{H}}_2OH$		$CH_3CH\underline{\mathbf{C}}H_3$	
$CH_3COO\underline{\mathbf{H}}$		$CH_3\underline{\mathbf{C}}HBrCH_3$	
$CH_3C\underline{\mathbf{H}}_2CH_3$		$CH_3\underline{\mathbf{C}}OOCH_3$	
$C\underline{\mathbf{H}}_3COCH_2CH_3$		$CH_3\underline{\mathbf{C}}OCH_3$	
$C_6\underline{\mathbf{H}}_6$		$CH_3\underline{\mathbf{C}}H(OH)CH_3$	
$CH_3C\underline{\mathbf{H}}O$		$CH_3{-}O{-}\underline{\mathbf{C}}H_3$	
$C\underline{\mathbf{H}}_3OH$		$H_2NCH(CH_3)\underline{\mathbf{C}}OOH$	
$CH_3COOC\underline{\mathbf{H}}_2CH_3$		$CH_3\underline{\mathbf{C}}ONHCH_3$	
$C_6H_5O\underline{\mathbf{H}}$		$H_2N\underline{\mathbf{C}}H(CH_3)COOH$	

3 **A ^{13}C NMR spectrum for each of the five compounds A to E (shown below) was obtained.**

A

B

C

D

E

The ^{13}C NMR spectra are numbered 1 to 5, and are shown below. Match up the spectra with the correct compound A to E. Explain how you deduced which is which. (AO2, AO3)

15 marks

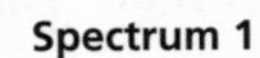

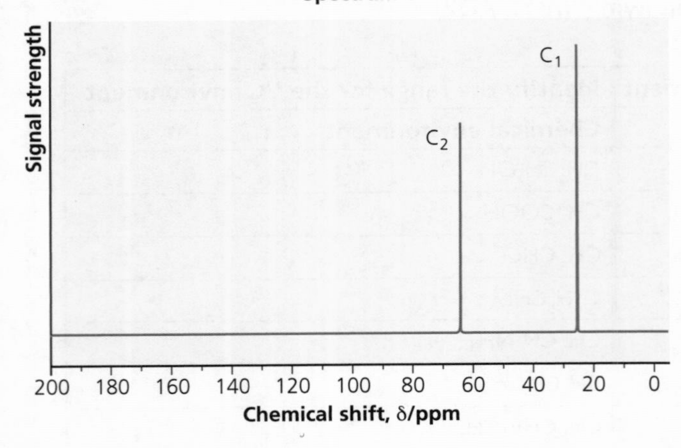

Spectrum 1 is compound .... because

....

....

Spectrum 2

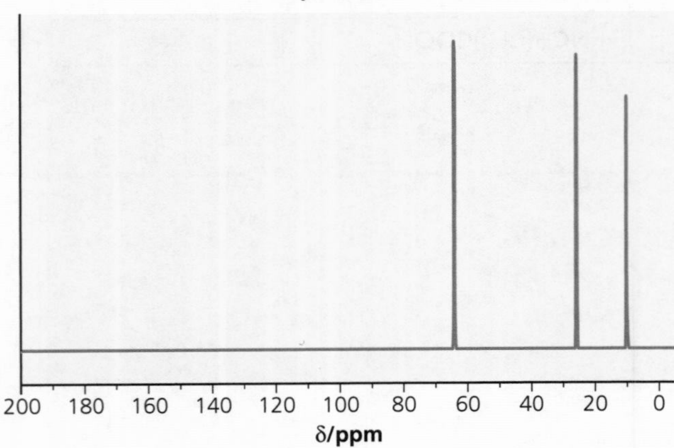

Spectrum 2 is compound .... because

....

....

Spectrum 3

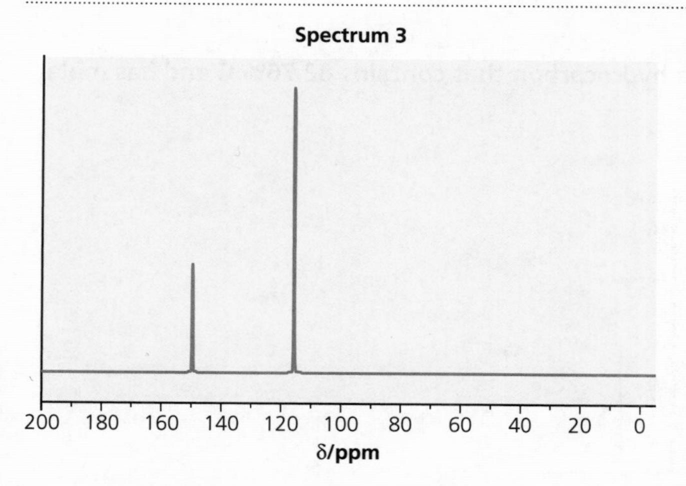

Spectrum 3 is compound .... because

....

....

Spectrum 4

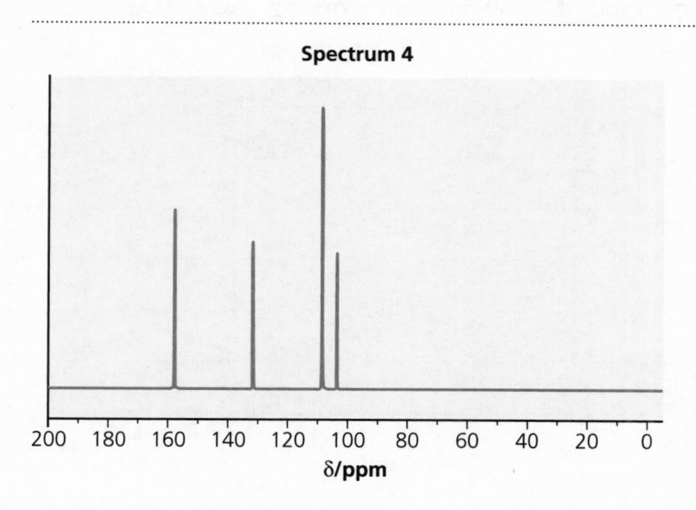

Spectrum 4 is compound .... because

....

....

Spectrum 5

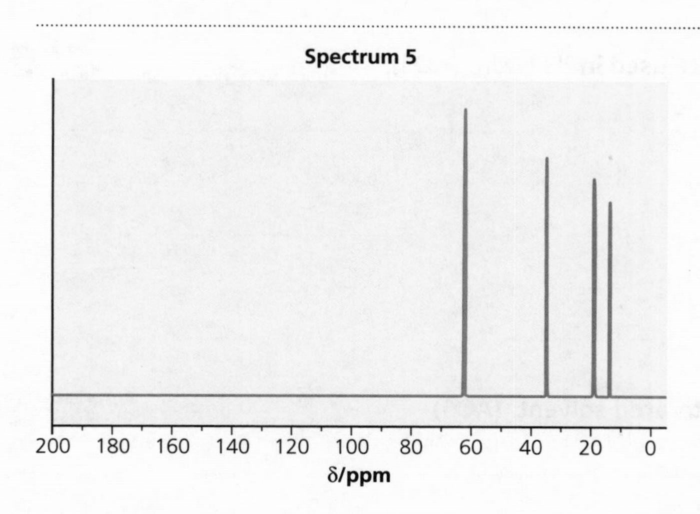

Spectrum 5 is compound .... because

....

....

4 The ^{13}C NMR spectrum below is that of a hydrocarbon that contains 82.76% C and has molar mass of $58\,g\,mol^{-1}$.

120 100 80 60 40 20 0
δ/ppm

Deduce the identity of the hydrocarbon compound. Show all your working. (AO2, AO3)

4 marks

....

....

....

....

....

....

....

....

....

....

....

5 a i Explain why deuterated solvents are used in ^{1}H NMR. (AO1)

2 marks

....

....

....

....

....

ii Give an example of a suitable deuterated solvent. (AO1)

1 marks

....

b **Use the '*n* + 1' rule to complete the table below. (AO1, AO2)** 2 marks

Number of H on adjacent C atoms	Type of splitting
1	
2	
3	
4	

c **Determine the relative peak areas in the ^{1}H NMR of 2-methylbutane. (AO2)** 2 marks

....

....

d **Complete the table below, part of which has been done for you. (AO2)** 6 marks

	CH_3CH_2OH			CH_3COOCH_3			$HCOOCH_2CH_3$		
Number of peaks	3								
Ratio of peak areas	3:2:1								
Type of H and δ value and splitting	C<u>H</u>$_3$	0.5–2.0	Triplet						
Type of H and δ value and splitting	C<u>H</u>$_2$	3.0–4.2	Quartet						
Type of H and δ value and splitting	O<u>H</u>	0.5–12	Singlet						

e **What is a labile proton? Explain how D_2O is used to help identify O–H and/or N–H protons. (AO1, AO3)** 3 marks

....

....

....

....

....

....

f **Identify *four* different functional groups that contain labile protons. (AO2)** 4 marks

....

....

Combined techniques

6 **In order to identify a compound, it is common to use a combination of more than one technique. Complete the following brief explanations about the use of different spectroscopic techniques. (AO1, AO2)**

a **Infrared spectroscopy can be used to identify** .... 1 mark

b **Chromatography can be used to** ....
in .... 2 marks

c **Mass spectrometry can be used to determine** .... 1 mark

d **In the GC-MS technique, GC is used to** ....
and MS uses .... **to match with a**
.... **to** .... 4 marks

e **^{13}C NMR is used to** .... 1 mark

f **^{1}H NMR is used to identify** ....
and .... **give information about**
.... 3 marks

7 **The spectra shown are for an unknown compound A. Use the hints given in the method/sequence outlined below the spectra to identify compound A. Show all of your working. (AO3)** 8 marks

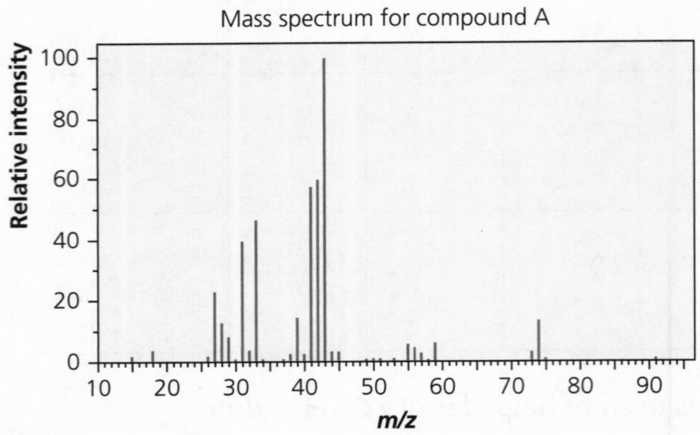

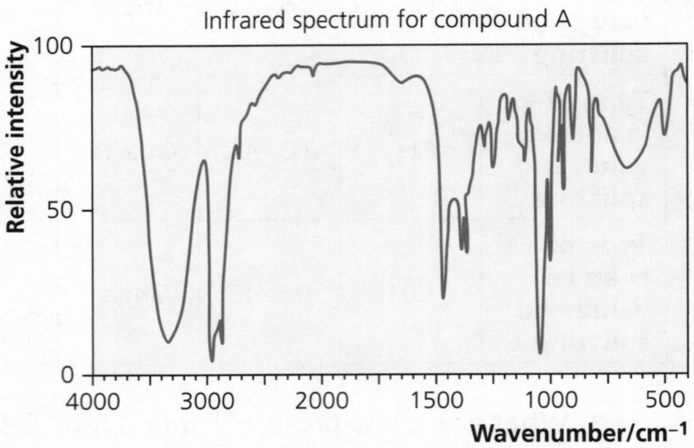

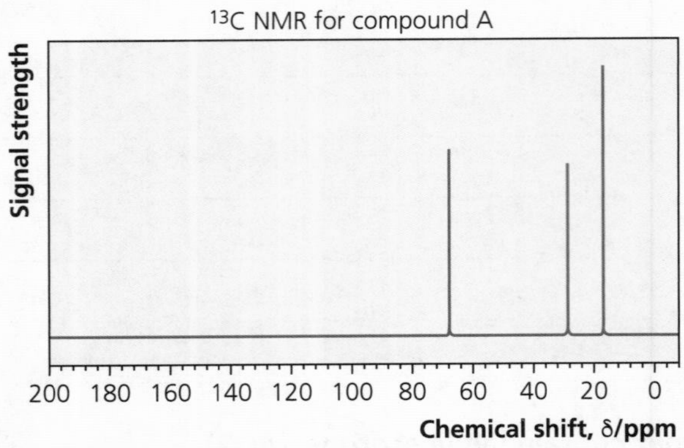

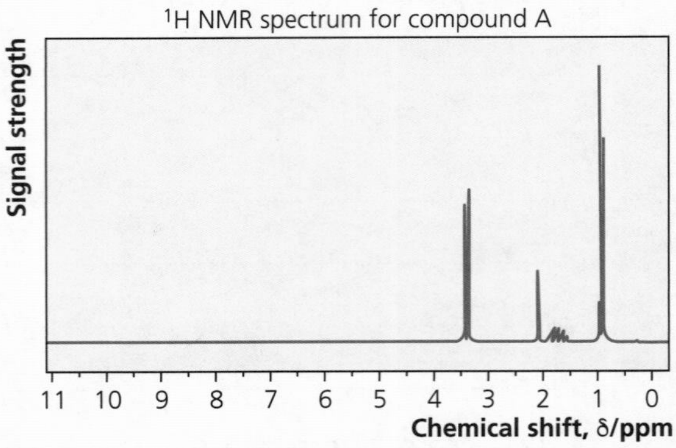

Mass spectrum for compound A

Relative intensity: 0, 20, 40, 60, 80, 100
m/z: 10, 20, 30, 40, 50, 60, 70, 80, 90

Infrared spectrum for compound A

Relative intensity: 0, 50, 100
Wavenumber/cm^{-1}: 4000, 3000, 2000, 1500, 1000, 500

^{13}C NMR for compound A

Signal strength
Chemical shift, δ/ppm: 200, 180, 160, 140, 120, 100, 80, 60, 40, 20, 0

^{1}H NMR spectrum for compound A

Signal strength
Chemical shift, δ/ppm: 11, 10, 9, 8, 7, 6, 5, 4, 3, 2, 1, 0

Method

Step 1. (Use IR to detect functional groups)

....

....

Step 2. (Use MS to determine molar mass)

....

....

Step 3. (Subtract mass of functional group from molar mass)

....

....

....

....

Step 4. (Deduce the numbers of atoms)

....

....

Step 5. (Use ^{13}C NMR to determine number of C environments)

....

....

Step 6. (Draw all possible isomers)

Step 7. **(Use ^{1}H NMR to determine number of H environments and use splitting patterns to deduce structure)**

....

....

....

....

Compound A is ....

8 **Identify compound B. Analyse the information to identify the compound. Include full details of your analysis of each of the spectra. Show all of your working. (AO3)** **7 marks**

Infrared spectrum for compound B

Transmittance/% — Wavenumber/cm^{-1}

Mass spectrum for compound B

Relative intensity — *m/z*

^{1}H NMR spectrum for compound B

Signal strength — Chemical shift, δ/ppm

^{13}C NMR for compound B

Signal strength — Chemical shift, δ/ppm

....

....

....

....

....

....

....

....

....

....

....

....

....

....

....

....

Compound B is ....

Exam-style questions

1 A chemist used gas chromatography to separate a mixture of esters.

a i How did the chemist use the results to predict the number of esters in the mixture and the amount of each ester. **1 mark**

....

....

ii Explain why the chemist was uncertain about the number of esters and the amount of each ester obtained from using gas chromatography alone. **2 marks**

....

....

....

....

b The chemist obtained a mass spectrum and both the ^{13}C and the ^{1}H NMR spectra of one of the esters. These are shown below.

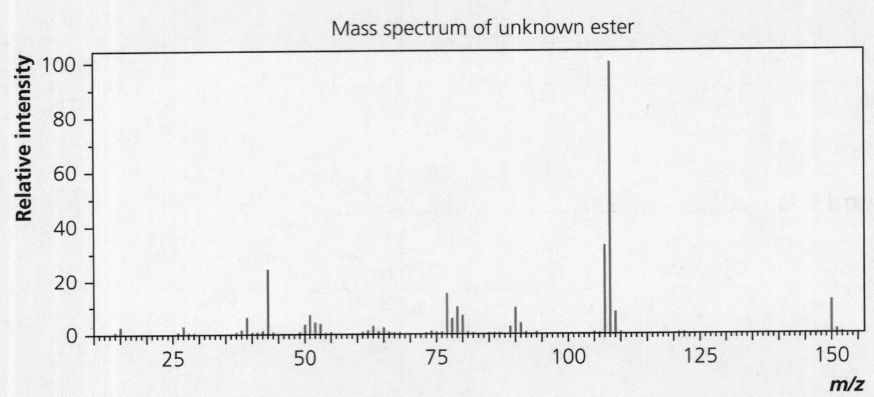

^{13}C NMR spectrum of unknown ester

131,130 127,126
122
169
149
16
21
200 180 160 140 120 100 80 60 40 20 0
δ/ppm

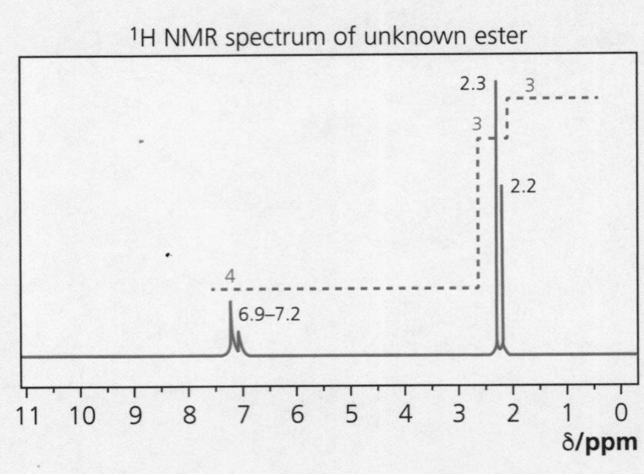

The chemist was able to narrow down the identity of the unknown ester to one of two compounds. Analyse the information to identify the two possible esters.

Include full details of your analysis of each of the spectra. **13 marks**

c Suggest how the spectra could be used to distinguish between the two esters.

2 a A food scientist carried out a series of chemical tests on an unknown compound X extracted from a sample of curry paste.

- The unknown compound X was added to a sample of bromine water, which immediately decolourised.
- The unknown compound X was refluxed with an acidified solution of potassium dichromate and the solution slowly turned dark green.

The food scientist used these observations to narrow down possible functional groups in the unknown compound X. Explain how. **3 marks**

b The food scientist then used the mass spectrum and the infrared spectrum (both shown below) to confirm the identity of the unknown compound X.

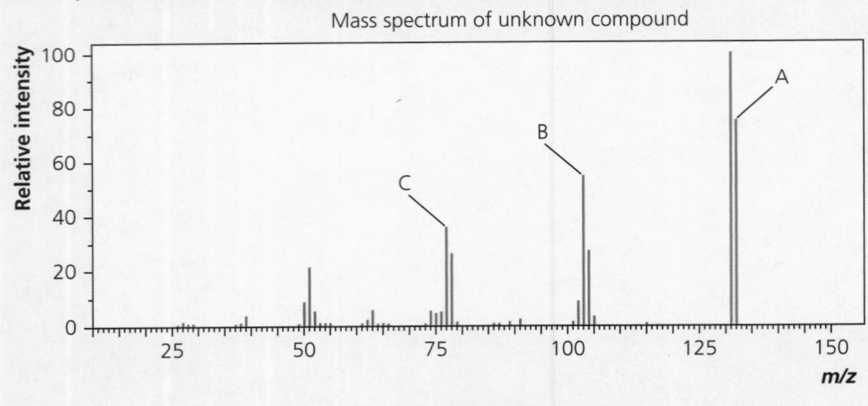

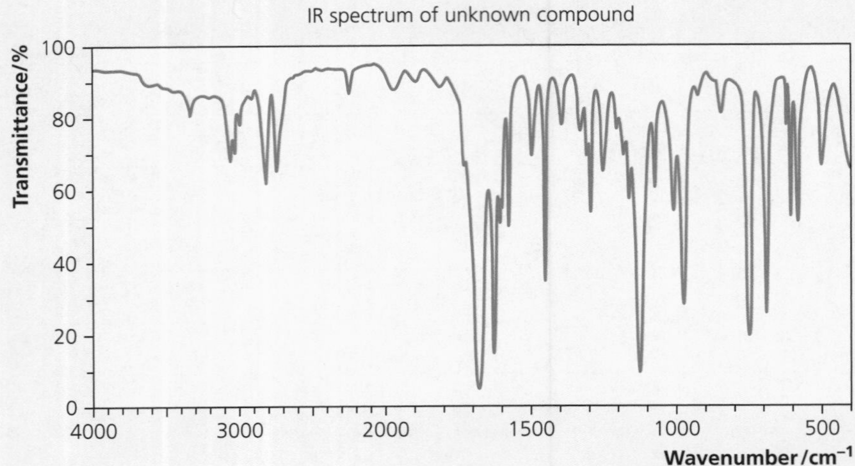

Analyse each of these spectra. Include full details of your analysis. Identify the ions responsible for peaks A, B and C in the mass spectrum. Use these peaks to identify the unknown compound.

10 marks

Also available

Philip Allan, an imprint of Hodder Education, an Hachette UK company, Blenheim Court, George Street, Banbury, Oxfordshire OX16 5BH

Orders

Bookpoint Ltd, 130 Park Drive, Milton Park, Abingdon, Oxfordshire OX14 4SE

tel: 01235 827827

fax: 01235 400401

e-mail: education@bookpoint.co.uk

Lines are open 9.00 a.m.–5.00 p.m., Monday to Saturday, with a 24-hour message answering service. You can also order through the Philip Allan website: www.philipallan.co.uk

ISBN 978-1-4718-4736-3

First printed 2016

Impression number 5 4 3 2 1

Year 2020 2019 2018 2017 2016

Cover photo reproduced by permission of Fotolia

Typeset by Aptara, Inc.

Printed in Spain

Hachette UK's policy is to use papers that are natural, renewable and recyclable products and made from wood grown in sustainable forests. The logging and manufacturing processes are expected to conform to the environmental regulations of the country of origin.